Fundamental Constants

Planck's constant:	$\hbar$	$=$	1.05457×10^{-34} J s
Speed of light:	c	$=$	2.99792×10^{8} m/s
Mass of electron:	m_e	$=$	9.10938×10^{-31} kg
Mass of proton:	m_p	$=$	1.67262×10^{-27} kg
Charge of proton:	e	$=$	1.60218×10^{-19} C
Charge of electron:	$-e$	$=$	-1.60218×10^{-19} C
Permittivity of space:	ϵ_0	$=$	8.85419×10^{-12} C^2/J m
Boltzmann constant:	k_B	$=$	1.38065×10^{-23} J/K

Hydrogen Atom

Fine structure constant:
$$\alpha = \frac{e^2}{4\pi\epsilon_0 \hbar c} = 1/137.036$$

Bohr radius:
$$a = \frac{4\pi\epsilon_0 \hbar^2}{m_e e^2} = \frac{\hbar}{\alpha m_e c} = 5.29177 \times 10^{-11} \text{ m}$$

Bohr energies:
$$E_n = -\frac{m_e e^4}{2(4\pi\epsilon_0)^2 \hbar^2 n^2} = \frac{E_1}{n^2} \quad (n = 1, 2, 3, \dots)$$

Binding energy:
$$-E_1 = \frac{\hbar^2}{2m_e a^2} = \frac{\alpha^2 m_e c^2}{2} = 13.6057 \text{ eV}$$

Ground state:
$$\psi_0 = \frac{1}{\sqrt{\pi a^3}} e^{-r/a}$$

Rydberg formula:
$$\frac{1}{\lambda} = R\left(\frac{1}{n_f^2} - \frac{1}{n_i^2}\right)$$

Rydberg constant:
$$R = -\frac{E_1}{2\pi\hbar c} = 1.09737 \times 10^7 \text{ /m}$$